Mathamusements

Raymond Blum

Illustrated by Jeff Sinclair

Sterling Publishing Co., Inc.
New York

Edited by Claire Bazinet

Library of Congress Cataloging-in-Publication Data

Blum, Raymond.
   Mathamusements / by Raymond Blum; illustrated by Jeff Sinclair.
      p.   cm.
   Includes index.
   Summary: An alphabetically arranged introduction to a variety
of math tricks and activities, including "Finger Math," "Geometric
Masterpieces," and "Numerology."
   ISBN 0-8069-9783-4
   1. Mathematical recreations—Juvenile literature. [1.
Mathematical recreations.] I. Sinclair, Jeff, ill. II. Title.
   QA95.B5185   1997
   793.7´4—dc21
                                                              96-49196
                                                                 CIP
                                                                  AC

10 9 8 7 6 5 4 3

Published by Sterling Publishing Company, Inc.
387 Park Avenue South, New York, N.Y. 10016
© 1997 by Raymond Blum
Illustrations © 1997 by Jeff Sinclair
Distributed in Canada by Sterling Publishing
c/o Canadian Manda Group, One Atlantic Avenue, Suite 105
Toronto, Ontario, Canada M6K 3E7
Distributed in Great Britain and Europe by Cassell PLC
Wellington House, 125 Strand, London WC2R 0BB, England
Distributed in Australia by Capricorn Link (Australia) Pty Ltd.
P.O. Box 6651, Baulkham Hills, Business Centre, NSW 2153,
Australia

Sterling ISBN 0-8069-9783-4 Trade
          ISBN 0-8069-9784-2 Paper

DEDICATION

This book is lovingly dedicated to my mother,
whose example taught me the importance
of working hard and always doing my best.

A very special thank-you to my daughter, Kate,
a talented writer and editor.

A NOTE TO PARENTS
AND TEACHERS

This book is filled with interesting and exciting math activities that will appeal to children of all abilities, ages nine and up. There are amazing number tricks, beautiful geometric designs, challenging puzzles, marvelous memory tricks, and many other mathematical amusements that will help spark children's interest in mathematics.

The book is organized so that children can open it up, pick out an activity, and get started on their own. The activities have clear, uncomplicated, step-by-step instructions, so that they are easy for children to read and understand. There is a glossary for looking up unfamiliar words. Any needed supplies can easily be found in the home or at school, or they can be purchased at minimal cost. Children will enjoy sharing these fascinating activities with their family and friends or their entire math class.

All of these activities have been classroom-tested, and children love them. Math teachers can use this book to supplement their math textbook to create interest and stimulate learning. When learning is fun and exciting, children become interested and are motivated to learn more. This book helps provide that motivation.

CONTENTS

ABRACADABRA

Magic tricks that use numbers are called mathe-magic tricks. This number trick is easy to learn and fun to perform for others. You will amaze your family and friends with your supernatural powers when you mysteriously reveal the number that has been secretly chosen!

What You Will Need

5 index cards   Pencil or marker

Preparing the Trick

First, copy the following numbers onto five index cards. Write the numbers exactly as shown here:

Card 1

1	3	5	7
9	11	13	15
17	19	21	23
25	27	29	31

Card 2

2	3	6	7
10	11	14	15
18	19	22	23
26	27	30	31

Card 3

4	5	6	7
12	13	14	15
20	21	22	23
28	29	30	31

Card 4

8	9	10	11
12	13	14	15
24	25	26	27
28	29	30	31

Card 5

16	17	18	19
20	21	22	23
24	25	26	27
28	29	30	31

Next, read the directions and practice the trick by yourself. When you have successfully worked it two or three times, you are ready to perform it for others.

What To Do

1. Ask your friend to secretly think of a number from 1 to 31.
2. Give her the 5 index cards and ask her to hand you each card that has her secret number on it.
3. As your friend hands you each card, glance at the number in the top left-hand corner.
4. Mentally add up these numbers. Their sum will be her secret number.

5. Finally, hold these cards to your forehead, close your eyes, and pretend that the cards are speaking to you as you reveal your friend's secret number!

Example

Suppose your friend chooses 19 as the secret number. She would hand you each card that has a 19 on it. The sum of the numbers in the top left-hand corner of each card (1 + 2 + 16) is 19.

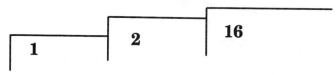

The Secret

This trick uses the binary number system that is based on the number 2. Look at the number in the top left-hand corner of each card: 1, 2, 4, 8, and 16. Each number has been multiplied by 2 to get the next number. They are called powers of 2. The rest of the numbers on each card have these powers of 2 as parts of their numbers.

For example, look at the card with the **8** in the top left-hand corner. The rest of the numbers on that card have **8** as part of their numbers:

$$10 = \mathbf{8} + 2$$
$$13 = \mathbf{8} + 4 + 1$$
$$28 = 16 + \mathbf{8} + 4$$
$$31 = 16 + \mathbf{8} + 4 + 2$$

Other Things To Do

Because the cards are numbered up to 31, you can have someone choose a secret number that is an important date, like an anniversary or a birthday.

BALANCING ACT

Look at the cartoon above and mentally draw a line straight down the middle. Both sides of the line are the same and the two sides of the cartoon are in balance. This is an example of symmetry. One half of the cartoon is the mirror image of the other half. Sometimes this is called mirror symmetry.

Paintblots

It is fun to make your own symmetrical artwork. It can be a little messy, but when you are finished you will have a beautiful symmetrical design.

What You Will Need

Newspaper
White paper

Poster paints
A small brush or stick

What To Do

1. Put down 2 or 3 layers of newspaper for a working surface.
2. Fold a piece of paper in half and crease it down the middle.
3. Open up the paper and lay it on the newspaper.
4. Use a brush or a stick to put blobs of poster paint on *one* side of the fold only. Do not put the blobs too close to the edges of the paper.
5. Fold the paper in half again and carefully press it closed.
6. Open up your symmetrical paintblot and lay it flat to dry.

Other Things To Do

Stringblots — Paint short pieces of string. Lay them on one side of the fold and press the paper closed. Open up the paper, remove the strings, and you will have a colorful symmetrical string design.

Artblots — Paint one half of a picture on one side of the fold. Use thick paint and make sure that your picture touches the crease. Press the paper closed. When you open it up you will have a symmetrical piece of art.

CRICKET THERMOMETER

Crickets are great little meteorologists. They "broadcast" the air temperature by their chirping. It is a scientific fact that temperature influences the chirp rate of crickets. They chirp slower when the air temperature is cooler and faster as the air gets warmer. So all you have to do is count cricket chirps to find the temperature of the air.

What You Will Need

Paper and pencil   A watch

What To Do

Use this Cricket Formula to find the air temperature in degrees Fahrenheit.

Temperature = number of chirps in 15 seconds + 40

Example

You count 32 chirps in 15 seconds. 32 + 40 = 72. The temperature would be about 72 degrees Fahrenheit.

Crickets may not be as accurate as a regular thermometer because this formula only gives you an estimate. That means that your answer is close to the actual temperature. However, this method of finding the air temperature is a lot more fun. Besides, even certified meteorologists are not 100% accurate!

Other Things To Do

1. To get a more accurate result, take several 15-second counts and average them. For example, you take 3 counts: 42 chirps, 39 chirps, and 45 chirps. To get their average, find their sum and divide by 3. 42 + 39 + 45 = 126 and 126 ÷ 3 = 42. Add 40 to this average and you get 82. The temperature would be about 82 degrees Fahrenheit.
2. Use this cricket formula to find the air temperature in degrees Celsius.

Temperature = (number of chirps in 15 seconds + 13) ÷ 2

For example, you count 40 chirps in 15 seconds. 40 + 13 = 53. Divide 53 by 2 and you get 26.5. The temperature would be about 27 degrees Celsius.

DAY YOU WERE BORN

Here is a riddle for you. How many birthdays does the average person have? The answer is 1 — the day that they were born!

Do you know on which day of the week you were born? Most people do not know and their parents usually cannot remember. By using an amazing math formula, you can find the day of the week of your birthday, or of any other important date from 1900–1999.

What You Will Need

Calculator   Paper and pencil

What To Do   **Example**
Born: Sept. 10, 1979

1. Write down the last two digits of the year you were born.   **79**

2. On a calculator, multiply that number by .25. Drop the decimal part of the number

if there is one. (.25 × 79 = 19.~~75~~)   **19**

3. Find the number of the month of
your birthday in the Months Table.
(September = 6)   **6**

4. Write the date of the month of your birthday. **+ 10**

5. Add your four answers from Steps 1 to 4.   **114**

6. *Without using a calculator,* divide
the sum you got in Step 5 by 7.
The remainder should be a number
from 0 to 6. It will be a 0 if 7 divides
into the sum exactly.

7. Find the *remainder* in the Days
Table. It will tell you on which day
of the week you were born.

$$\begin{array}{r} 16 \\ 7\overline{)114} \\ \underline{7} \\ 44 \\ \underline{42} \\ \longrightarrow \underline{\mathbf{2}} \end{array}$$

2 = Monday

MONTHS TABLE			
January	1*	July	0
February	4**	August	3
March	4	September	6
April	0	October	1
May	2	November	4
June	5	December	6

* Use **0** in leap years
** Use **3** in leap years

DAYS TABLE	
Sunday	1
Monday	2
Tuesday	3
Wednesday	4
Thursday	5
Friday	6
Saturday	0

Now ask your parents if they remember on which
day of the week they were born. Then use the formula to check and see how good their memory is.

EGYPTIAN PYRAMIDS

If you enjoy puzzles, this mind bender is the one for you. It is an exciting puzzle that is fun as well as challenging. See if you can discover the secret strategy so that you can build your very own Egyptian Pyramid.

What You Will Need

A deck of playing cards   3 index cards

What To Do

1. Place 3 index cards next to each other on a table. Label them A, B, and C.
2. Remove the Ace (1), 2 and 3 of Diamonds from the deck.
3. Make a three-card pyramid with the cards face

up. A pyramid is a geometrical figure that is larger at the bottom than it is at the top. So, put the 3 of Diamonds at the bottom, the 2 in the middle, and the Ace (1) at the top.

4. Place the pyramid on top of pile A.

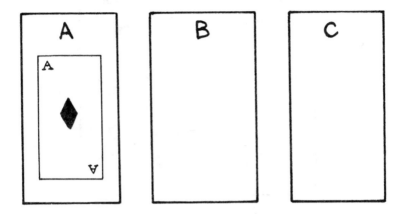

How To Move

1. The object of the puzzle is to move the entire three-card pyramid from pile A to pile C *in the least number of moves*.

2. Move only one card at a time.

3. Each move consists of moving one card from the top of one pile to the top of either of the other two piles.

4. You may not put a larger-numbered card on top of a smaller-numbered card. It is okay to put a 2 on top of a 3, but you cannot put a 3 on top of a 2.

You should be able to move the three-card pyramid in just 7 moves. If it takes you more than 7 moves, start over and try a different strategy. You will find the solution in the answer section at the back of this book.

The Secret

Here is how to figure out the least number of moves for different-size pyramids.

Number of Cards	Least Number of Moves
1	1
2	3
3	7
4	15
5	31
6	63
7	127
8	255

Do you see the pattern in the second column? If you multiply each number by 2 and then add 1, you get the next number.

$1 \times 2 = 2$ and $2 + 1 = \mathbf{3}$
$3 \times 2 = 6$ and $6 + 1 = \mathbf{7}$
$7 \times 2 = 14$ and $14 + 1 = \mathbf{15},$ and so on.

Other Things To Do

Make the puzzle harder by making a taller pyramid. Then see if you can move that pyramid in the least number of moves. If you have a lot of spare time, try moving a thirteen-card pyramid (Ace through King). It will only take you 8,191 moves!

FINGER MATH

Do you know that you can use your fingers to multiply? If you ever forget some of your multiplication facts, you can just use your fingers to find the answer. It is easy, fast, and fun, too. Your family and friends will be amazed when you show them how to multiply on your "hand calculator"!

What To Do

1. Hold your hands, palms up, in front of you. Each finger has the value shown.

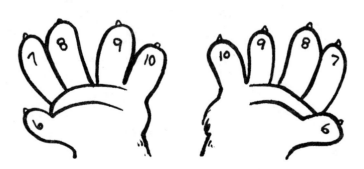

2. To multiply two numbers, touch those two fingers together.

Example 1   **7 × 8**

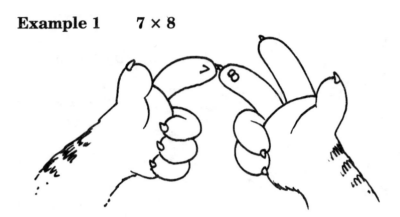

3. Add the fingers that are up to get the number in the tens place.
4. Multiply the fingers that are down to get the number in the ones place.

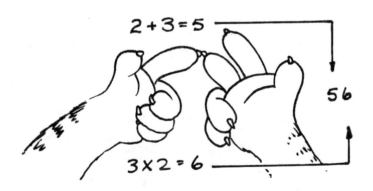

So 7 × 8 = 56

Example 2   6 × 10

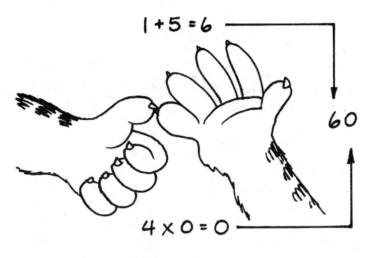

So 6 × 10 = 60

Exceptions: 6 × 6, 6 × 7, and 7 × 6. When you multiply the fingers that are down, you get a number that is greater than 9. So, for the exceptions, you have to carry a 1 and add it to the number in the tens place.

Example 3   6 × 7

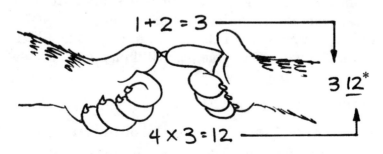

* The number in the ones place is greater than 9, so carry the 1 and add it to the 3. The final answer is 42.

GEOMETRIC MASTERPIECES

Geometry is a kind of mathematics that studies points, lines, angles, and different shapes. You can create many interesting geometric designs by using only a compass and a ruler. Then, if you add a little color, you will have some beautiful pieces of geometric art to display!

What You Will Need

An 8½ in × 11 in (21.5 cm × 28 cm) piece of plain white paper
A compass and ruler

Pencil and eraser
Colored pencils or markers (optional)

What To Do

1. Open your compass about 3½ inches (9 cm). This will be the circle's radius (the distance from the center to any point on the circle).
2. Find the center of your paper and draw a circle.
3. Lightly mark a point A at the top of the circle. Then, *without changing the compass setting*, place the compass point on point A and lightly mark a point B on the circle.

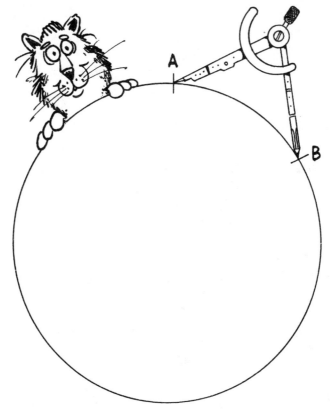

4. Place the compass point on point B and lightly mark a point C. Continue in this way around the

circle until you have marked 6 points (A–F). Your circle should now be divided into six parts.

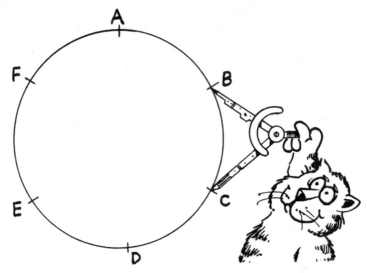

5. Use your ruler to copy the basic pattern on your circle. You will have to erase some of these lines later so *draw all of your lines very lightly*.

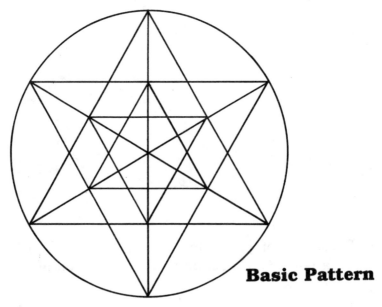

Basic Pattern

6. If you like the basic pattern just as it is, go to step 7. If you want to copy one of the examples, just erase those lines in your basic pattern that are not part of the example.

Examples

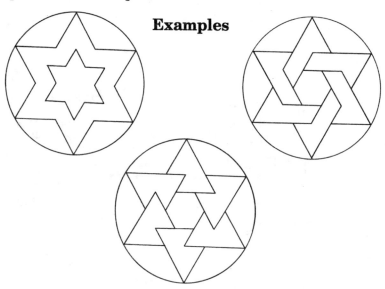

7. Color your design with colored pencils or markers.

Other Things To Do

1. Start with the basic pattern and then erase some lines or add more lines to create an original geometrical design.

2. Make circle patterns. Follow steps 1 through 4 but only open your compass 2 inches (5 cm). Then, *without changing the setting of the compass*, draw 6 new circles using the 6 points for the centers.

Happy 1,000,000th BIRTHDAY

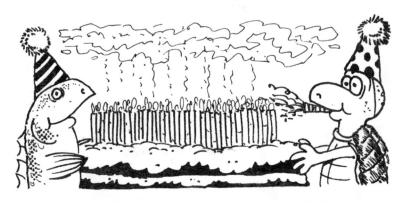

Do you know how old you are in seconds? Have you reached your one million or one billion seconds birthday yet? Maybe you have a "seconds birthday" coming up soon and you do not even know about it!

What You Will Need

Paper and pencil   A calculator

TIME CHART			
1 minute	=	60	seconds
1 hour	=	60	minutes
1 day	=	24	hours
1 year	=	365.25	days

How long is *one million (1,000,000)* seconds?

What To Do

1. Enter 1,000,000 in your calculator.
2. To change the number of seconds to minutes, divide by 60.

3. To change the number of minutes to hours, divide your result by 60.

4. To change the number of hours to days, divide your result by 24.

Did you get about 11.57 or approximately 12 days?

Everyone probably missed your one million seconds birthday, but there is still hope. Check and see if your one billion seconds birthday is coming up soon.

How long is *one billion (1,000,000,000)* seconds?

What To Do

1. One billion is one *thousand* million. So multiply the number of days you got for one million seconds (11.57) by 1000. This will give you the number of days in 1 billion seconds.

2. To change the number of days to years, divide your result by 365.25

There is a Seconds Birthday Chart in the back of the book so that you can check your answer. While you are there, check and see if you have any other "seconds birthdays" coming up soon.

Other Things To Do

You can figure out your approximate age in seconds. Multiply your approximate age × 365.25 × 24 × 60 × 60.   (Use .5 for ½)

Example

Age ≈ 11½ years old (≈ means approximately equal to)
11.5 × 365.25 × 24 × 60 × 60 ≈ 362,912,400 seconds! (On your calculator, multiphy 11.5 × 365.25 × 24 × 6 × 6 and then add 2 zeros to your answer.)

IMPOSSIBLE KNOT

Can you pick up one end of a piece of rope in each hand and tie a knot in the rope *without letting go of the ends*? Impossible! Or is it? Actually it is easy to do, but first you have to know the mathemagical secret!

What You Will Need

A long piece of string, cord, or rope
   (about 30 in/75 cm)

What To Do

1. Place the piece of rope on a table.

2. Tie a knot in your arms by crossing them.

3. Bend over the piece of rope and pick up one end of the rope in each hand.

4. Straighten up and carefully uncross your arms without letting go of the ends. The knot that was in your arms has been transferred to the piece of rope!

The Secret

This trick uses a special kind of mathematics called topology. Topology is the study of shapes and what happens to those shapes when they are folded, pulled, bent, or stretched out of shape. Topology also helps us perform amazing tricks that appear to be impossible!

Other Things To Do

Try this topological trick on your family and friends. Wait until they get themselves all tied up in knots before you show them the mathemagical secret.

JOURNEY TO THE MOON

If someone wants to lose weight fast, they should take a trip to the moon. Weight is the result of the pull of gravity. The moon's gravitational pull is much less than that of the Earth's. As a result, a person who weighs 110 pounds (50 kg) on Earth weighs only about 18 pounds (8 kg) on the moon! To find about how much you would weigh on the moon, just divide your Earth weight by 6. You can also figure out how much you would weigh on each of the other planets in our solar system.

What You Will Need

A calculator   Paper and pencil

What To Do

All the planets in our solar system have different gravitational pulls. As a result, you would weigh more on some planets and less on others. To find your approximate weight on each of the planets, just multiply your Earth weight by the gravity factor.

WEIGHT CHART

Place	Gravity Factor
Mercury	.38
Venus	.88
Earth	1.00
Mars	.38
Jupiter	2.53
Saturn	1.19
Uranus	.91
Neptune	1.13
Pluto	.06

Example

Your weight on Earth is 88 pounds (40 kg). Approximately how much would you weigh on the surface of Jupiter?

$88 \times 2.53 \approx 223$ pounds $(40 \times 2.53 \approx 101$ kg)

Other Things To Do

**My Very Educated Mother Just Showed
Us Nine Planets!**

If you memorize the sentence above, you can easily remember the order of the planets from the sun. The first letter of each word helps us remember the first letter of each of the planets in our solar system.

**Mercury Venus Earth Mars Jupiter Saturn
Uranus Neptune Pluto**

KNUCKLE MONTHS

THE DAYS OF THE MONTHS

Thirty days hath September,
April, June, and November.
February has twenty-eight alone,
All the rest have thirty-one,
Excepting Leap Year—that's the time
When February's days are twenty-nine.

Many people memorize a little poem like this, then recall it when they need to know whether a month has 30 or 31 days. It works fine if you can remember the poem, but there is an easier way. Just look at your knuckles!

What To Do

1. Hold your hands in front of you, palms down, and then make two fists.
2. When you make two fists, you can see your eight top knuckles and the spaces between them. From left

34

to right, these knuckles and spaces represent the months of the year.

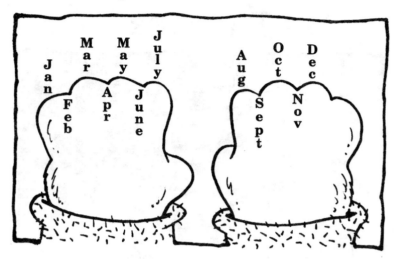

3. Find the month that you are looking for and see how many days it has.

The months on the knuckles have 31 days (January, March, May, July, August, October, and December).

The months in the spaces between the knuckles are those that have 30 days (April, June, September, and November).

February, as usual, is the only exception. It is in a space between the knuckles but it has only 28 days (29 days in a leap year).

You can memorize the little poem or use your knuckles to remember the number of days in each month. The knuckle method is better. If you don't think so, ask someone to recite "The Days of the Months" poem. Most adults just stop at the end of second line to figure out what months they left out! They don't even remember the whole poem.

LIGHTNING AND THUNDER

You see a huge flash of lightning in the sky. Then, BOOM, you hear the rumble and roar of thunder. How far away is the storm? There is an easy way to calculate it. All you have to do is count.

What You Will Need

Paper and pencil

What To Do

1. Go inside where it is safe to observe the storm.
2. Start counting seconds when you see a flash of

lightning. Count slowly to yourself saying, "One, one thousand; two, one thousand; three, one thousand," and so on.

3. Stop counting when you hear the crash of thunder.

4. Divide the number of seconds you get by 5 (for miles) or by 3 (for kilometers). This will tell you approximately how far away you are from the storm.

Examples

You count 15 seconds. 15 ÷ 5 = 3 so the storm is about 3 miles away. (15 ÷ 3 = 5 km)

You count 3 seconds. 3 ÷ 5 = .6 so the storm is about .6 mile away. (3 ÷ 3 = 1 km)

The Secret

Light travels very fast. The speed of light is about 186,000 miles per second (300,000 km/sec). Sound travels much more slowly. The speed of sound is about 1,100 feet per second (330 m/sec).

So even though lightning and thunder start at the same place, we see the lightning almost immediately but the sound of the thunder takes longer to reach our ears. Sound travels about 1 mile in 5 seconds or about 1 kilometer in 3 seconds.

Other Things To Do

You can tell if a storm is going away from you or coming toward you by repeating your count every few minutes. If you count more seconds, the storm is moving away from you. If you count fewer seconds, the storm is approaching you.

MR. OWL ATE MY METAL WORM

What is so special about the word radar, the phrase "never odd or even," and the number 747? They are all palindromes. A palindrome is any group of letters or numbers that reads the same forward and backward. The title of this section is an example of a sentence that is a palindrome. Any two-digit number that is not a palindrome can be turned into one by just reversing the number and adding. Some numbers take a little longer than others, but eventually a palindrome will magically appear. Watch out, though, for the dreaded number 98. You will need a little time and a sharp pencil to find its palindrome!

What You Will Need

Paper and pencil   A calculator

What To Do

Example

1. Pick any two-digit number that is not
a palindrome.   **49**
2. Reverse the number and add it to
the first number.   **+94**
                                           14 3

3. Continue reversing the sums and
adding until you get a palindrome.   **+341**
                                           484

The number 49 becomes a palindrome after two additions.

Other Examples

$$
\begin{array}{r}
8\,6 \\
+\,6\,8 \\
\hline
1\,5\,4 \\
+\,4\,5\,1 \\
\hline
6\,0\,5 \\
+\,5\,0\,6 \\
\hline
1\,1\,1\,1
\end{array}
$$

three additions ➛ 1 1 1 1

$$
\begin{array}{r}
6\,9 \\
+\,9\,6 \\
\hline
1\,6\,5 \\
+\,5\,6\,1 \\
\hline
7\,2\,6 \\
+\,6\,2\,7 \\
\hline
1\,3\,5\,3 \\
+\,3\,5\,3\,1 \\
\hline
4\,8\,8\,4
\end{array}
$$

four additions ➛ 4 8 8 4

Are you ready to try 98? The number 89 is the same. Both of these numbers turn into palindromes after 24 additions! Sharpen your pencil and add carefully. Then, see if your palindrome matches the answer in the back of the book. (Hint: After 12 additions, your sum will be 85,189,247!)

Other Things To Do

Start with a three-digit or a four-digit number. Most of them will eventually turn into a palindrome after a few additions. For example, the number 381 turns into a palindrome after 4 additions. If you want a greater challenge, try 739. It becomes a palindrome after 17 additions. You can check your answer in the back of the book. Number 196 is an exception, however. Mathematicians have used computers to reverse and add it *thousands of times*, but its palindrome has still not appeared!

NUMEROLOGY

In the science of numerology everyone is assigned a number. Can this special number possibly reveal some important information about your personality? Try it yourself and see if it works.

What You Will Need

Paper and pencil

What To Do

1. Print your name vertically (up and down). Use your full name as it appears on your birth certificate. Do not use your nickname. Include your middle name if you have one.

2. Next to each letter, write its number from the Letter Values chart below.

LETTER VALUES

1	2	3	4	5	6	7	8	9
A	B	C	D	E	F	G	H	I
J	K	L	M	N	O	P	Q	R
S	T	U	V	W	X	Y	Z	

3. Add up all the numbers from your name.
4. If your answer has more than one digit, add the digits together until there is only one digit. This one-digit number is your name personality number.
5. Look at the Personality Portrait and see what your name personality number says about you.

Example   THOMAS PATRICK JOHNSON

T	2		P	7		J	1
H	8		A	1		O	6
O	6		T	2		H	8
M	4		R	9		N	5
A	1		I	9		S	1
+ S	1		C	3		O	6
			+ K	2		+ N	5
	2 2			3 3			3 2

22 + 33 + 32 = 87

   87 ⟶ 8 + 7 = 15, and 15 ⟶ 1 + 5 = 6

His name personality number is 6.

Other Things To Do

Find your birth date personality number. Then compare it to your name personality number and see which one is closer to who you really are.

Example
   Thomas's birth date is **October 14, 1987**
                    or   **(10-14-1987)**

   10 + 14 + 1987 = 2,011

   2,011⟶ 2 + 0 + 1 + 1 = 4

His birth date personality number is 4.

PERSONALITY PORTRAIT

1. There is no one else like you. You are a natural leader. Originality and independence are two of your strengths. You have the courage to try new things.

2. You are kind and gentle. People say that you are considerate and sensitive to the feelings of others. You are the best friend anyone could have. Writing poetry comes naturally to you.

3. You are the life of the party. Center stage is where you feel right at home. You have a great sense of humor. Your true talents are found in art and writing.

4. People can count on you to get things done. You always work hard to achieve your goals. Endless patience and self-discipline are your strengths. You are very loyal to your friends.

5. You are very intelligent. Traveling to far-off places interests you. You are very curious and love to investigate new things. Risk taking is one of your strengths.

6. You have strong moral values. People know that they can trust you and that you are honest. You set the example for everyone else. Helping and caring for others are two of your strong points.

7. You are very smart for your age. Solving mysteries interests you. You have a strong love of nature and animals. You are a thinker and like to spend time by yourself.

8. You have the power to succeed and you are always organized. Management and authority are two of your strengths. Making money comes naturally to you. An executive position is in your future.

9. Current events interest you. You are always working to help everyone else. Human rights is something that you are concerned about. Kindness and understanding are two of your strong points.

ORIGAMI

Origami is a Japanese word that means "the folding of paper." If you follow these simple step-by-step instructions, you will turn a piece of plain paper into the shape of a cube. This origami figure is also known as a water bomb, balloon, or paper ball.

What You Will Need

An 8½ in × 11 in (21.5 cm × 28 cm) sheet of paper

Scissors

What To Do

1.Take a plain sheet of paper. Fold up the bottom left-hand corner, as shown below.

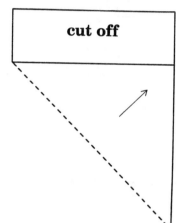

cut off

2. Crease the paper on the dotted line. Then cut off the extra paper at the top and you will have a square.

3. Fold and crease on the dotted lines. Do this two or three times back and forth to get flexible creases.

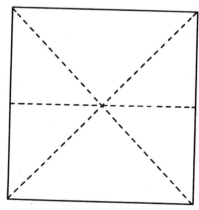

4. Fold down the top half.

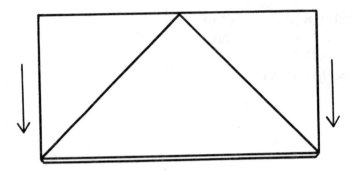

5. Push in the top corners to the center.

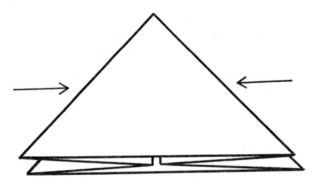

Do steps 6–8 on the front side of the shape first.

6. Fold up the bottom corners to the top and crease.

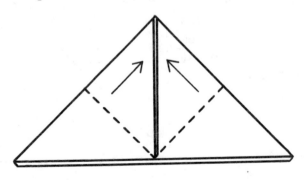

7. Fold in the side corners to the center and crease.

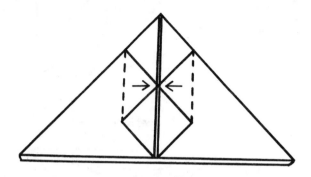

8. Fold down the two small top flaps and crease. Tuck them into the two center pockets.

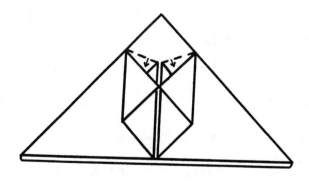

9. Turn the paper over and repeat steps 6–8 on the back side of the shape.

10. Blow into the small hole at the bottom to inflate the cube.

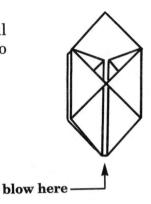

blow here

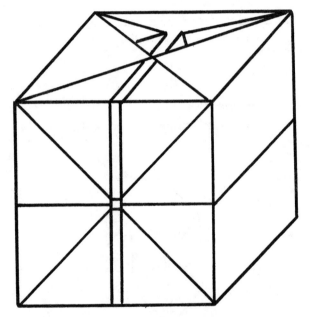

11. Carefully crease any edges that have not already been creased.

Other Things To Do

Your cube is very well constructed if it holds water. Make a little funnel out of another piece of paper and then carefully fill your cube. Good luck!

PIZZA PI

Pi (pronounced "pie") = 3.141592653
5897932384626433832795028841971 6
9399375105820974944592307816406 2
8620899862803482534211706798214 8
0865132823066470938446095505822 3
1725359408128481117450284102701 9
3852110555964462294895493038196 4
4288109756659334461284756482337 8
67831652712019091456 48...

What does this extremely long number have to do
with pizza? Keep reading and you will find out!

What You Will Need

Paper and pencil   A calculator

What To Do

1. The following chart contains the approximate measurements of the circumferences and diameters of five objects. Circumference is the distance around a circle and diameter is the distance across the center of a circle. Use your calculator to divide each circumference by its diameter and then fill in the last column.

Object	Circum-ference	Diameter	circumference / diameter
Coin	116 mm	37 mm	_____
Plate	85 cm	27 cm	_____
Table	3.83 m	1.22 m	_____
Pizza*	117.4 m	37.4 m	_____
Earth**	40,075 km	12,757 km	_____

2. Check your answers in the back of the book. Do you see that each answer is a little more than 3? The answers are not the same because the measurements are not exact. For every circle in the universe, no matter what its size, the number you get by dividing the circumference by the diameter is always the same. This number is called pi and is

*The largest pizza ever baked was one measuring 385 feet 4 inches in circumference and 122 feet 8 inches in diameter.

symbolized by the Greek letter π. The ancient Greeks worked out pi to four decimal places — 3.1416. Today mathematicians have used computers to calculate pi to billions of decimal places. Pi is usually rounded to 3.14 when doing calculations.

Pi is an amazing number because its decimal never ends and its digits never repeat in any pattern. It just keeps on going and going forever and ever. The first 255 decimal places of pi are in the first paragraph of this section.

Other Things To Do

Wrap a piece of string around any circular object (bike wheel, soda can, jar lid, spool of thread, wastebasket). Measure the piece of string with a ruler to find the circle's circumference. Then measure the diameter of the circle. Divide the circumference by the diameter and see how close your answer is to pi.

**The circumference of the Earth is about 24,902 miles and its diameter is about 7,927 miles.

QUICKLY, WHAT COMES NEXT?

What comes after 157? The answer is obviously 158. Most people would have no problem answering that question. However, when you ask your friends a simple "what comes next" question, they will get the answer wrong almost every time!

What To Do

Tell your friend that you are going to *quickly* read ten numbers. After each number, your friend is to *immediately* say the next higher number. For example, if you say "twenty-eight," your friend should say "twenty-nine."

Here is the list of numbers:

seventy-two

one hundred twenty-eight

thirteen

five

eight hundred fifty-six

two thousand eight hundred sixty-five

seven hundred eighty-one

thirty-four

five hundred seventy-three

four thousand ninety-nine

Did your friend answer "five thousand" after the last number? The next number after 4,099 is not 5,000. It is 4,100! Tell your friend not to feel too badly because most people answer incorrectly. Be sure to read this list to your family too, and see how good they are at taking a simple math quiz!

The Secret

You are not giving your friend much time to think, and it is very natural to say five thousand after you hear the number four thousand in 4,099.

Other Things To Do

What comes next in the sequence of letters below?

O, T, T, F, F, S, S, ___, ___

Check your answer in the back of the book.

Hint: The answer is as easy as 1, 2, 3!

A **R**AT IN THE HOUSE

MIGHT EAT THE
ICE CREAM

Look at the first letter of each word in the title above. Put the letters together and what do they spell? A-R-I-T-H-M-E-T-I-C !

This is an example of a memory trick called a mnemonic (ni mon ik—the *m* at the start is silent). A mnemonic is a word, phrase, rhyme, or anything used to help you remember. So if you can remember the silly phrase about the ice cream–eating rats, you can remember how to spell the word arithmetic.

Here are a few more mnemonics that will help you remember some things you will most likely need to know for your math class.

Is Less Than (<), Is Greater Than (>)

The symbols are formed by your two hands. Most people use their left hand *less* and their right hand *more* (greater).

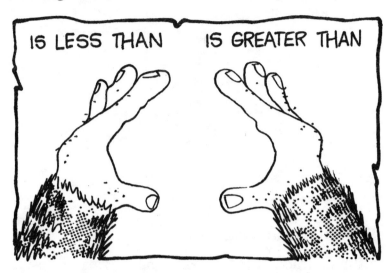

How Many Teaspoons in a Tablespoon?

Both teaspoon and tablespoon start with the letter t. What number rhymes with t?...3!

3 teaspoons = 1 tablespoon

The Order of Operations in Long Division

Example 79 ÷ 3

<u>D</u>ad	<u>D</u>ivide	<u>D</u>ivide 7 by 3.	**26**
<u>M</u>om	<u>M</u>ultiply	<u>M</u>ultiply 2 by 3.	3)79
<u>S</u>ister	<u>S</u>ubtract	<u>S</u>ubtract 6 from 7.	**6**
<u>B</u>rother	<u>B</u>ring down	<u>B</u>ring down the 9.	**19**
<u>R</u>over	<u>R</u>estart or	<u>R</u>estart — Divide 19	**18**
	<u>R</u>emainder	by 3, and so on.	**1**

The Clockwise Order of Compass Points

<u>N</u>ever <u>E</u>at <u>S</u>hredded <u>W</u>heat, or
<u>N</u>ever <u>E</u>at <u>S</u>oggy <u>W</u>affles, or
<u>N</u>ever <u>E</u>at <u>S</u>limy <u>W</u>orms, or
<u>N</u>ever <u>E</u>at Septic Waste

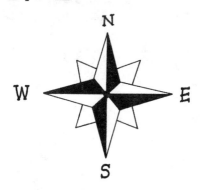

Dividing Fractions

"If it's a fraction you are dividing by, turn it upside down and multiply."

Example

$$\frac{3}{8} \div \frac{1}{2} \qquad \frac{1}{2} \text{ turned upside down is } \frac{2}{1}$$

$$\text{So } \frac{3}{8} \div \frac{1}{2} = \frac{3}{8} \times \frac{2}{1} = \frac{6}{8} = \frac{3}{4}$$

Perimeter

<u>Peri</u>meter is the distance around the <u>rim</u> (border, edge, boundary) of a surface or figure.

The Distance Formula

<u>d</u> <u>i</u> <u>r</u> <u>t</u>

⟶ <u>d</u>istance <u>is</u> equal to <u>r</u>ate × <u>t</u>ime

Pi (π) Rounded To Ten Decimal Places (3.1415926536)

The number of letters in each word reveals each digit.

May I have a large container of orange juice
3 . 1   4   1   5   9   2   6   5

now please ?
3   6

The Order of Operations   Example

Please   Parentheses $20 \div 5 + \underline{(6 - 4)} \times 3^2$

Excuse   Exponents $20 \div 5 + 2 \times \underline{3^2}$

My Dear   Multiplications
              and Divisions $\underline{20 \div 5} + \underline{2 \times 9}$

Aunt Sally Additions and
              Subtractions $\underline{4 + 18}$
                                  22

Trigonometric Ratios

$\text{sine} = \dfrac{\text{opposite}}{\text{hypotenuse}}$ ⟵─────── "soh-cah-toa"

$\text{cosine} = \dfrac{\text{adjacent}}{\text{hypotenuse}}$

$\text{tangent} = \dfrac{\text{opposite}}{\text{adjacent}}$

Multiplying Two Binomials

FOIL Multiply the First terms, the Outside terms, the Inside terms, and the Last terms

Example

$$\begin{array}{cccc} & F & O & I & L \end{array}$$
$(x + 3)(x + 4) = x \cdot x + 4 \cdot x + 3 \cdot x + 3 \cdot 4 =$
$x^2 + 4x + 3x + 12 = x^2 + 7x + 12$

SPELLING BEE

M-A-T-H-A-M-U-S-E-M-E-N-T-S!

This mathemagical card trick is easy to learn and fun to perform, too. You'll amaze family and friends when you spell the names of different playing cards and those cards suddenly and mysteriously appear!

What You Will Need

A deck of playing cards

Preparing The Trick

1. Remove all 13 Hearts from the deck.
2. Arrange the Hearts in the following order (it is called the setup):

Q - 4 - A - 8 - J - 2 - 7 - 5 - 10 - K - 3 - 6 - 9

3. Put the Hearts back in the deck *in this order*. Disguise the setup by spacing these cards throughout the whole deck. Put the 9 of Hearts near the bottom of the deck and the Queen of Hearts near the top.
4. Read the directions below and then practice the trick by yourself. When you have successfully worked the trick two or three times, you are ready to perform it for others.

Performing The Trick

1. Starting at the bottom of the deck, carefully remove all the Hearts. Remove the 9 first and put it *face down* on the table. Then remove the 6 and put it *face down* on top of the 9. Continue in this way until you have removed all 13 Hearts. Make sure that you keep the Hearts *in the same order* as the setup. Put the rest of the deck aside.

2. Show your friend the pack of Hearts. Say that you are going to make each card magically appear by just spelling the card's name.

3. Hold the pack of Hearts face down in your hand. Say "O" as you put the top card *at the bottom of the pack*. Say "N" as you put the next top card at the bottom of the pack. Say "E" and then turn over the next top card. It will be a ONE (Ace)! Remove that card from the pack and put it on the table.

4. Use the same method to remove the rest of the Hearts from the pack. (Spell ELEVEN for Jack, TWELVE for Queen, and THIRTEEN for King.) Remember, when you say a letter, put the top card at the bottom of the pack. When you say the *last* letter of each word, turn over the top card and it will be the card that you just spelled!

The Secret

The setup is the secret to doing this trick. The Hearts must be arranged in exactly that order or the trick will not work.

Other Things To Do

When you are setting up the deck, arrange all the Clubs, too. Then, if your friend wants you to repeat the trick, you will be ready.

TANGRAM PUZZLES

The tangram puzzle is a Chinese invention that has been enjoyed by people around the world for thousands of years. It consists of seven puzzle pieces called "tans" that can be put together to make hundreds of different shapes and figures. See if you can solve the mystery of this ancient Chinese puzzle.

TANGRAM PUZZLE

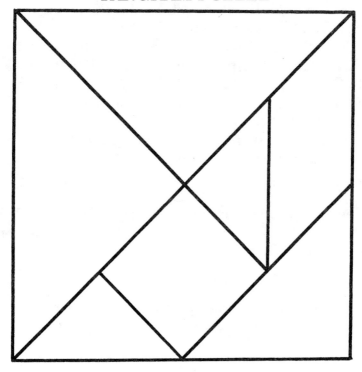

What You Will Need

A piece of plain white paper and a pencil
A ruler

Scissors
Glue stick
An empty cereal box

What To Do

1. Photocopy the square tangram puzzle or use your ruler to *carefully* trace it onto a piece of white paper.
2. Glue that copy of the puzzle onto the side of a used cereal box.
3. *Carefully* cut along the black lines to make the seven tans.
4. Mix up the tans. Then see if you can put the square back together again. Good luck!
5. Try to make the following geometric shapes. No pieces can overlap and you must use *all seven tans*.

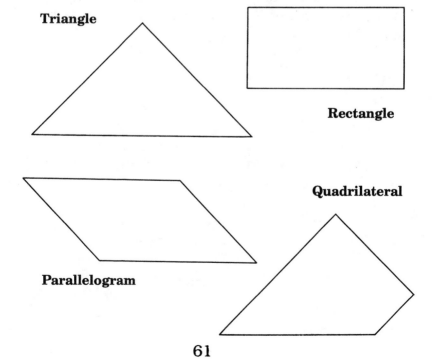

Triangle

Rectangle

Parallelogram

Quadrilateral

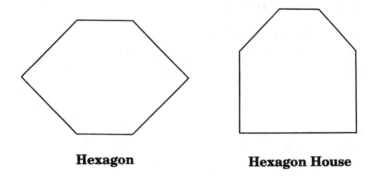

Hexagon   **Hexagon House**

Check your solutions in the back of the book.

The Secret

The puzzles are a little easier to solve if you understand the basic relationship between the different tans. Look at your set of tangram pieces. The area of the medium triangle is half the area of a large triangle. The area of the two small triangles together equals the area of the medium triangle, or the area of the parallelogram, or the area of the square.

Other Things To Do

1. Check out a tangram puzzle book from the library. It will contain shapes of animals, people, houses, boats, birds, and much more.
2. Create your own tangram designs and see if your friends can put them together.

UNBELIEVABLE MAGIC

This is an unbelievable magic trick that you can perform for your family and friends. It is an easy trick to learn and it will leave everyone completely baffled.

What You Will Need

A flat washer with a diameter of at least 1⅜ in (3.5 cm) with a large hole in the center

A 55-in (140 cm) piece of string

Preparing the Trick

1. Tie the two ends of the string together. Snip off the extra little pieces of string at the end of the knot.
2. Read the directions below and then practice the trick with someone in your family. (Or sit on the floor and stretch the string between your two big toes!) When you have successfully worked the trick two or three times, you are ready to perform it for others.

What To Do

1. Slip the string through the washer and stretch it between your friend's thumbs. Then, tell your friend that you are going to remove the washer from the string without removing the string from his thumbs—and without removing his arms from his body!

2. Pinch the string at point **A** with your right hand. Pinch the string at point **B** with your left hand.

3. Pull your right hand toward you and push your left hand away from you.

4. Loop the string that is in your left hand over your

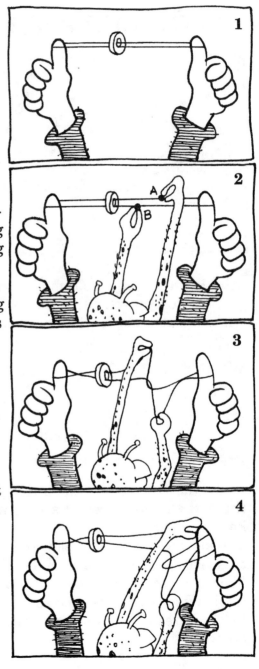

friend's left thumb and let go. Do not let go of the right-hand pinch.

5. With your left hand, pinch the string at point C.

6. Loop the string that is in your left hand over your friend's left thumb as in step 4. Your friend will have to move his thumbs a little closer together so that you can do this. Again, do not let go of the right-hand pinch.

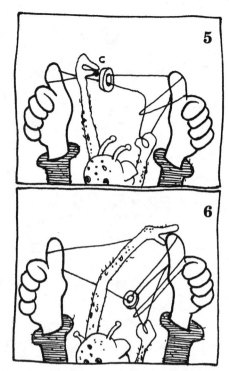

Then, ask your friend to pinch each thumb and index finger together so that the string will not slide off over the tops of his thumbs. Finally, release the right-hand pinch and tell your friend to *slowly* pull his hands apart. Unbelievably, the washer will fall to the ground and the string will still be looped around your friend's thumbs!

The Secret

You take the string off your friend's thumb with the first loop. The second loop puts the string back on his thumb *on the other side of the washer*. So, when the string is stretched out, it stays on his thumb but the washer slides off the end. Just as in the Impossible Knot, topology helps you perform an amazing trick that at first seems impossible.

VERY FAST MULTIPLYING

WHAT IS 75 × 75 ?!.. 5,625 !!!

You can amaze your family and friends by multiplying large numbers in your head. It is easy to do when you know the secret shortcuts!

Multiplying a two-digit number by 11

Example   26 × 11

What To Do

1. Separate the two digits   **2 _ 6**
2. Add the two digits together.   **2 + 6 = 8**
3. Put that sum between the two digits.   **2 8 6**
   So 26 × 11 = 286

If the sum of the two digits is greater than 9, you have to carry a 1 and add it to the first digit.

Example   84 × 11

(8 + 4 = 12)   **8 12* 4**

*The sum of the digits is greater than 9, so carry the 1 and add it to the 8. The final answer is **924**.

The Secret

Multiply 26 × 11 to see why this trick works.

```
        2 6
    ×  1 1
      ─────
        2 6
      2 6
      ─────
```

first digit ⟶ 2 8 6 ⟵ second digit

2 + 6 = 8

Squaring a two-digit number ending in the number 5

("Squaring" a number means multiplying a number by itself.)

Example   75 × 75

What To Do

1. Multiply the first digit by one more than itself. (One more than **7** is **8**)   **7 × 8 = 56**

2. Put **25** after that answer (the answer will always end in 25).   **5625**

So 75 × 75 = 5625

The Secret

This trick uses an algebraic procedure called squaring a binomial, and only works when the number you are squaring ends in 5.

WALK THROUGH A PIECE OF PAPER

Do you know that you can cut a hole in a piece of notebook paper big enough for you to walk through? Does it sound impossible? It's not! You just need to know the mathemagical secret.

What You Will Need

Scissors
A ruler

An 8 1/2 in × 11 in (21.5 cm × 28 cm) piece of paper

What To Do

1. Fold a piece of paper in half from top to bottom.
2. Start cutting along the folded edge. Make 8 parallel cuts about 1 inch (2.5 cm) apart. Do

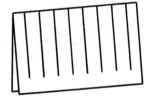

not cut all the way to the open edge. Stop cutting about 1 inch (2.5 cm) from the edge of the paper.

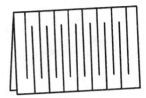

3. Now start cutting along the open edge. Make 7 parallel cuts in the middle of the other 8 cuts. Again, stop cutting about 1 inch (2.5 cm) from the edge of the paper.

4. Carefully unfold the paper and then flatten it out. Cut through all the folds *except the two end sections*.

5. If you *carefully* stretch out the paper, you should be able to walk right through it!

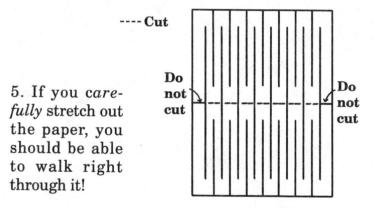

The Secret

It's topology at work again. As in Impossible Knot and Unbelievable Magic, topology has helped you solve a problem that at first seemed impossible.

Other Things To Do

Use a larger piece of paper and make more cuts using the same pattern. Make sure that the number of cuts along the folded edge is one more than the number of cuts along the open edge. Maybe you can make a hole large enough for you and your friend to walk through!

X-TRA SPECIAL DESIGNS

Do you know that you can make curved lines by drawing straight lines? Line designs are drawn with a straight ruler, but many of the lines look curved. These fascinating geometrical designs are easy to draw and a lot of fun, too!

What You Will Need

Plain white paper and a pencil
Colored pencils or markers

A ruler
An eraser

What To Do

1. Photocopy the blank line designs or use your ruler to *carefully* trace each one onto a piece of white paper.
2. Follow the directions for each design. Connect the points using your ruler and pencil. Use colored pencils or markers to make the most colorful drawings.
3. Check the completed design opposite, or in the back of the book.

Directions

Plus Sign — In each of the four sections, draw straight lines between the two points that have the same number.

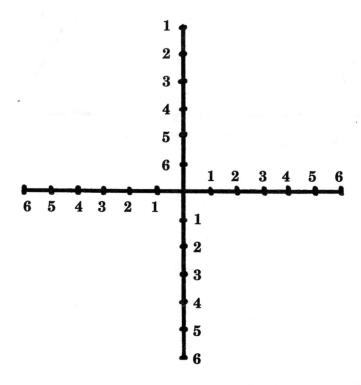

Triangle — Make triangles by drawing straight lines between the three points that have the same number.

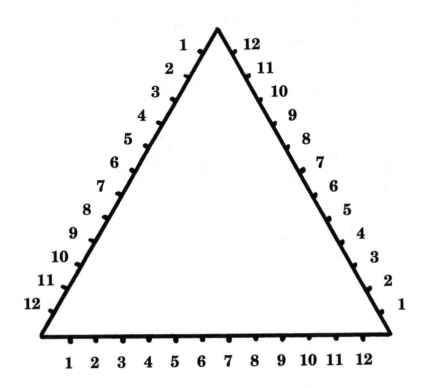

Square — Make squares by drawing straight lines between the four points that have the same number.

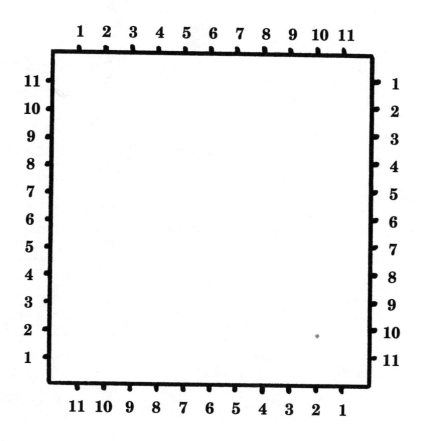

Circle — Draw a straight line from each odd-numbered point on the circle to each even-numbered point on the circle. For example, connect 1 to 2, 1 to 4, 1 to 6, and so on until you get back to where you started. Then connect 3 to 4, 3 to 6, 3 to 8, and so on until you get back to where you started. Continue around the circle until you finish point 11.

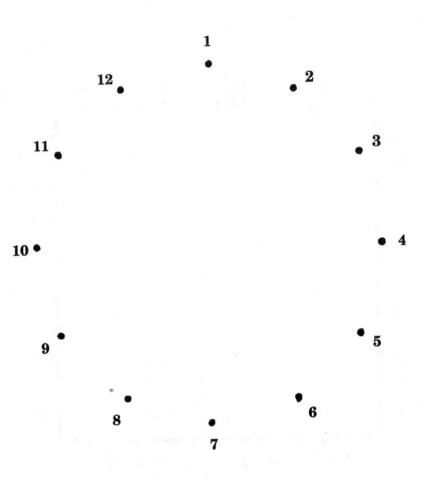

The Secret

All of these geometric line designs are constructed by following a mathematical pattern. This gives the illusion of curved lines even though all the lines are straight.

Other Things To Do

You can make larger designs using the same mathematical patterns.

75

YOUR FAVORITE NUMBER

Do you know that your calculator is an electronic magician? Just push the right buttons and your calculator will perform mathemagic for you.

Here are two "favorite number" calculator tricks that are sure to amaze your family and friends.

What You Will Need

A calculator

Trick #1

What To Do

1. Hand your friend a calculator and ask her to enter the number **37,037**.
2. Ask her for her favorite number from 1 to 9.
3. Mentally multiply her favorite number by 3.
4. Tell her to multiply the number in the calculator by that answer.
5. Your friend will end up with six of her favorite numbers all in a row!

Example

7

$(7 \times 3 = 21)$

37037 × 21

777777

The Secret

Multiply 37,037 × 3 on your calculator and you will uncover the secret.

Trick #2

What To Do

	Example
1. Hand your friend a calculator and ask him to enter his favorite 3-digit number.	**123**
2. Tell him to enter the number again so that he now has a 6-digit number.	**123123**
3. Ask him to divide that number by 7, then by 11, and finally by 13.	
4. He will end up with his favorite 3-digit number!	**123**

The Secret

Entering a 3-digit number twice is the same as multiplying the 3-digit number by 1,001. Dividing by 7, 11, and 13 is the same as dividing by 1,001. One operation cancels out the other and you end up with the 3-digit number that you started with!

Other Things To Do

Pick a favorite number from 1 to 9. Multiply that number by 12,345,679. (Notice that the 8 is missing.) Then multiply that answer by 9. Use paper and pencil to multiply by 9 if your answer does not fit in your calculator. You will end up with a whole row of your favorite numbers! Multiply 12,345,679 × 9 and you will see why this trick works. You can check your answers in the back of the book.

ZOOMING

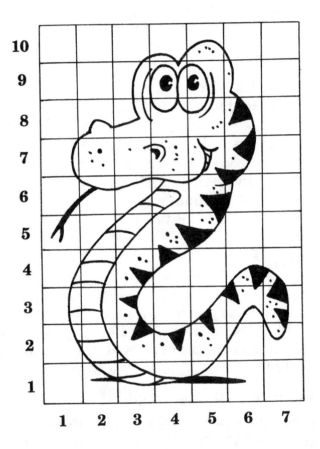

A zoom lens is used to enlarge the image that is seen in a camera. A small distant object appears closer and larger when the lens is zoomed in. You can get the same result using only paper and pencil. The secret is a checkerboard pattern called a grid.

What You Will Need

An 8½ in × 11 in (21.5 cm × 28 cm)
   piece of plain white paper
Colored pencils or markers

A ruler
A pencil and
   eraser

What To Do

Make the Enlarged Grid

Tip: Pencil-in the lines of your grid lightly so that you will be able to erase them easily.

1. Use your ruler to draw a 7 × 10 inch (17.5 cm × 25 cm) rectangle on a piece of white paper. The sides of the rectangle should be parallel to the sides of the paper and the corners must be square (at right angles). Make a mark every inch (2.5 cm) along the four sides of the rectangle, as shown in figure 1.

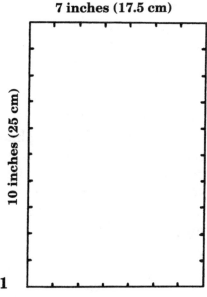

Figure 1

2. Connect the marks on the opposite sides of the rectangle with parallel lines Number the left side and the bottom of the grid as shown on the next page, figure 2.

Figure 2

Enlarged Grid

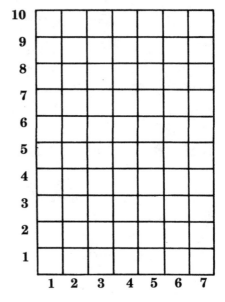

Enlarge the Cartoon

1. Transfer the cartoon to the enlarged grid one section at a time. Use the numbers on the left side and at the bottom of the grid to help you.
2. Color the enlarged cartoon with colored pencils or markers.
3. Erase all the grid lines.

The Secret

The enlarged cartoon is similar to the original one. Its size is larger but its shape and proportions stay the same. The sides of each section of the enlarged grid are about 2.5 times as long as the sides of each section of the smaller grid. So the area of the enlarged cartoon is about six times the area of the original cartoon.

Other Things To Do

1. Use an extra large piece of paper and a yard (meter) stick and you can make your enlargement as big as you want.
2. Draw small grids (or place clear plastic grid sheets) over other favorite cartoons or artwork and—zoom! —you can quickly decorate your whole room!

ANSWERS

Egyptian Pyramids

Three-Card Pyramid

Moves	A	B	C
start	1	–	–
1.	2	–	1
2.	3	2	1
3.	3	1	–
4.	–	1	3
5.	1	2	3
6.	1	–	2
7.	–	–	1

Happy 1,000,000th Birthday

SECONDS BIRTHDAY CHART

Name	No. of Seconds	Approx.Time
one	1	1 sec.
thousand	1,000	17 min.
million	1,000,000	12 days
billion	1,000,000,000	32 yrs.
trillion	1,000,000,000,000	32,000 yrs.
quadrillion	1,000,000,000,000,000	32,000,000 yrs.
quintillion	1,000,000,000,000,000,000	32,000,000,000 yrs.

(Continue to multiply by 1,000 to get the next higher number.)

Mr. Owl Ate My Metal Worm

98

After 24 additions, 98 will turn into the palindrome 8,813,200,023,188.

739

After 17 additions, 739 will turn into the palindrome 5,233,333,325.

Pizza Pi

Object	circumference / diameter
Coin	3.1351351
Plate	3.1481481
Table	3.1393443
Pizza	3.1390374
Earth	3.1414126

Quickly, What Comes Next?

One
  Two
    Three
      Four
        Five
          Six
            Seven ⟶ Eight, Nine

Tangram Puzzles

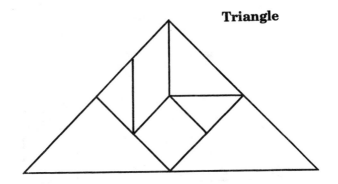

Triangle

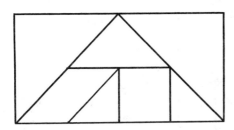

Rectangle

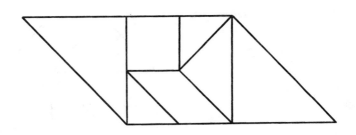

Parallelogram

Quadrilateral

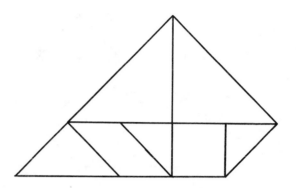

Hexagon

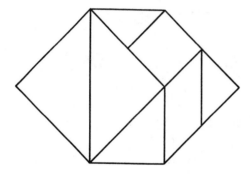

Hexagon House

84

X-tra Special Designs

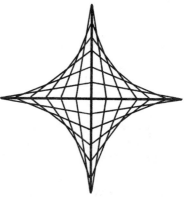

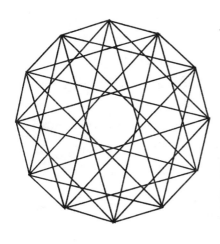

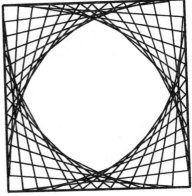

Your Favorite Number

$12,345,679 \times 1 \times 9 = 111,111,111$
$12,345,679 \times 2 \times 9 = 222,222,222$
$12,345,679 \times 3 \times 9 = 333,333,333$
$12,345,679 \times 4 \times 9 = 444,444,444$
$12,345,679 \times 5 \times 9 = 555,555,555$
$12,345,679 \times 6 \times 9 = 666,666,666$
$12,345,679 \times 7 \times 9 = 777,777,777$
$12,345,679 \times 8 \times 9 = 888,888,888$
$12,345,679 \times 9 \times 9 = 999,999,999$

GLOSSARY

algebra A mathematical language that uses letters along with numbers. $5x + 6 = 21$ is an example of an algebra problem.

approximately equal to ($\approx$) A symbol used when an answer is close to an exact answer.

area The amount of space inside a figure.

average The sum of a set of numbers divided by how many numbers there are.

billion A word name for 1,000,000,000.

binary number system A number system based on the number 2.

binomial An algebraic expression that has two terms. Example: $2x + 1$

birth date The date of the day that you were born.

Celsius The temperature scale of the metric system.

center of a circle The point that is the same distance from all of the points on a circle.

centimeter A metric unit of length that approximately equals .4 of an inch.

circumference The distance around a circle.

circumference of the Earth Approximately 24,902 miles (40,075 km).

compass An instrument used to draw circles.

compass points North, East, South, and West.

cube A three-dimensional figure with six square faces all the same size.

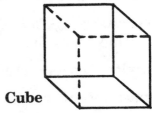

Cube

day A unit of time equal to 24 hours.

decimal part of a number The digits to the right of the decimal point.

diameter The distance across the center of a circle.

diameter of the Earth Approximately 7,927 miles (12,757 km).

digit Any of the symbols 0 to 9 used to write numbers. Example: 6,593 is a four-digit number.

distance formula distance = rate × time

estimate To give an approximate rather than an exact answer.

even numbers The numbers 0, 2, 4, 6, 8, 10....

Fahrenheit The temperature scale of the U.S. system.

formula An algebraic sentence that states a math

fact or rule. Example: The area of a rectangle equals the length times the width. (A = l w).

geometric Consisting of straight lines, circles, angles, triangles, etc.

geometry A kind of mathematics that studies points, lines, angles, and different shapes.

gravity The force that pulls things downward.

gravity factor The number that you multiply your Earth weight by to find your approximate weight at different places in our solar system.

grid Horizontal and vertical parallel lines in a checkerboard pattern.

hexagon A polygon with six sides.

horizontal line A line that runs straight across from left to right.

hour A unit of time equal to 60 minutes.

hundreds place Example: In the number 8,376, the 3 is in the hundreds place.

inch A U.S. unit of length equal to 2.54 centimeters.

is greater than (>) A symbol used to compare two numbers when the larger number is written first. Example: 73 > 5

is less than (<) A symbol used to compare two numbers when the smaller number is written first. Example: 12 < 47

kilogram A metric unit of mass that approximately equals 2.2 pounds.

kilometer A metric unit of length that approximately equals .6 of a mile.

leap year A year having 366 days. A leap year is a year that can be divided by 4 exactly. Examples: 1996, 1992, 1988, 1984, 1980, etc.

light (speed of) Approximately 186,000 miles per second (300,000 km/sec).

line design A geometric design made with straight lines.

mathemagic Magic tricks that use numbers.

meteorologist A person who studies and reports the weather.

mile A U.S. unit of length that approximately equals 1.6 kilometers.

million A word name for 1,000,000.

minute A unit of time equal to 60 seconds.

mirror symmetry When one half of a figure is the mirror image of the other half.

mnemonic A word, phrase, rhyme, or anything that can be used to help you remember.

numerology Assigns everyone a number based on their name or birth date. This number might reveal information about your personality.

odd numbers The numbers 1, 3, 5, 7, 9, 11....

ones place Example: In the number 8,376, the 6 is in the ones place.

operations +, -, ×, and ÷

order of operations Rules about the order in which operations should be done. 1. Parentheses 2. Exponents 3. From left to right, multiplications and divisions 4. From left to right, additions and subtractions

origami A Japanese word that means "the folding of paper."

palindrome Any group of letters or numbers that reads the same forward and backward.

parallel lines Lines in the same plane that never intersect.

parallelogram A quadrilateral with two pairs of parallel sides.

perimeter The distance around the rim or border of a polygon.

pi (π) The number obtained by dividing the circumference of a circle by its diameter. It approximately equals 3.14.

polygon A closed two-dimensional figure with three or more sides.

powers of 2 Each number is multiplied by 2 to get the next number. 1,2,4,8,16....

pyramid A three-dimensional figure whose base is a polygon and whose faces are triangles with a common vertex.

Pyramid

quadrilateral A polygon with four sides.

quadrillion A word name for a 1 with 15 zeroes after it: 1,000,000,000,000,000.

quintillion A word name for a 1 with 18 zeroes after it: 1,000,000,000,000,000,000.

radius The distance from the center of the circle to any point on the circle.

rate The speed of an object.

rectangle A parallelogram with four right angles.

remainder The number left over after dividing.

repeating decimal A decimal in which a digit or group of digits to the right of the decimal point repeats forever. Example: 17.333333333....

right angle An angle that has a measure of 90 degrees.

sequence A set of numbers in a certain pattern or order. Example: 3, 6, 9, 12....

setup When cards or props are arranged before performing a magic trick.

similar figures Figures that have the same shape but may not have the same size.

sound (speed of) Approximately 1,100 feet per second (330 m/sec).

square A parallelogram with four right angles and four equal sides.

squaring a number When a number is multiplied by itself. Example: $7 \times 7 = 49$

sum The answer to an addition problem.

symmetry What a shape has when it can be folded in half and the two halves match exactly.

tablespoon A U.S. unit of measure equal to three teaspoons.

tangram puzzle A seven-piece puzzle that can be put together to make hundreds of different shapes and figures.

tans The seven puzzle pieces of a tangram puzzle.

teaspoon A U.S. unit of measure equal to ⅓ of a tablespoon.

tens place Example: In the number 8,376, the 7 is in the tens place.

thousand A word name for 1,000.

thousands place Example: In the number 8,376, the 8 is in the thousands place.

topology A kind of mathematics that studies shapes and what happens to those shapes when they are folded, pulled, bent, or stretched out of shape.

triangle A polygon with three sides.

trillion A word name for 1,000,000,000,000.

vertex (plural: vertices) The point where lines meet to form an angle.

vertical line A line that runs straight up and down.

year A unit of time equal to 365¼ or 365.25 days.

Metric Conversion Chart

Into Metric

You know number of:	Multiply by (times):	To get number of:
inches	2.54	centimeters
feet	30.5	centimeters
yards	0.91	meters
miles	1.6	kilometers

Out of Metric

You know number of:	Multiply by (times):	To get number of:
millimeters	0.04	inches
centimeters	0.4	inches
meters	3.3	feet
kilometers	0.62	miles

INDEX

ABOUT THE AUTHOR

RAYMOND BLUM has been a mathematics teacher for over 25 years. In 1991, he wrote the book *Mathe-magic*, which is filled with dozens of entertaining number magic tricks. His second book, *Math Tricks, Puzzles & Games*, was published in 1994. It is a fascinating collection of fun-filled math activities. Both books are written for children ages nine and up and for teachers to use in their classrooms.

Ray Blum has been a speaker at numerous state and national math conferences, where he shares his classroom magic with other teachers. For years, he has performed a number magic show for elementary and middle school children as "Professor Numbers." The professor shows children the magical, fun side of mathematics with his mathemagic and arithme-tricks. Ray has won several teaching awards, including 1994 Wisconsin Teacher of the Year, and is currently teaching math at James Madison Memorial High School in Madison, Wisconsin.

ABOUT THE ARTIST

JEFF SINCLAIR has been drawing cartoons ever since he could hold a pen. He has won several local and national awards for cartooning and humorous illustration. When he is not working away at his drawing board, Jeff can be found renovating his home and feeding hungry koi in his backyard water garden. Jeff has recently gone into cyberspace on the Internet. He currently lives in Vancouver, BC, Canada with his wife, Karen; son, Brennan; daughter, Conner; and Golden Lab, Molly.